C

Electricity For A Developing World: New Directions

Christopher Flavin

D1292365

Worldwatch Paper 70
June 1986

©Copyright Worldwatch Institute, 1986
Library of Congress Catalog Card Number 86-50567
ISBN 0-916468-71-2

Table of Contents

Electric power systems, long viewed as showpieces of development, are now central to some of the most serious problems Third World countries face. Many Third World utilities are so deeply in debt that international bail-outs may be necessary to stave off bankruptcy. Financial troubles, in conjunction with various technical problems, have led to a serious decline in the reliability of many Third World power systems—which may impede industrial growth. The common presumption that developing countries will soon attain the reliable, economical electricity service taken for granted in industrial countries is now in doubt.

Electric power is considered an essential infrastructure investment, similar to roads and water supply systems that underpin economic development. Since the late fifties, electric power has been a priority of government planners and international development agencies alike. Budgets for electrification have grown rapidly and total spending now approaches $50 billion each year. The World Bank and the regional development banks have devoted about 20 percent of their total lending to electricity development and encouraged commercial banks to invest heavily in Third World power systems. Power projects now account on average for about one quarter of public capital investments in developing countries, or up to 2 percent of gross national product.[1]

I would like to thank Douglas Barnes, Paul Clark, James Fish, Howard Geller, Nigel Green, Alan Miller, and David Zoellner for their useful reviews of a draft version of this paper.

6 The electric-generating capacity of the Third World in the mid-eighties is approximately 450,000 megawatts (including China, which by itself has a generating capacity of 86,000 megawatts). This is just two thirds of the capacity of the United States alone which had 688,000 megawatts in 1985. The Third World has just 120 watts of electricity per person, compared to 2,900 watts per person in the United States—a 24-fold difference. However, electricity consumption in most developing countries is growing at the rapid pace of between 5 and 15 percent annually. The burgeoning economies of Brazil and South Korea, for example, are projected to nearly triple their power use during the next 15 years.[2]

Although electricity decisions are couched in the cold language of input-output models and strategic planning, they influence the basic outlines of societies that are still being shaped. Electric power decisions can affect the balance of power among industries, cities, and regions, and between urban and rural areas. Bringing electricity to a remote village is one of the most fundamental changes that can alter rural life. In short, questions about the nature and scope of electricity programs affect some of the most basic issues facing the Third World today, including the balance between the public and private sectors and the widening gap between rich and poor.

Between 90 and 95 percent of the electric power investments in developing countries goes to providing power to large cities and industries.[3] Planners explain this as a logical priority given the needs of the modern sector and the wider benefits that are expected to result from industrialization. In some countries, such as the neo-industrial economies of the Far East, this traditional approach to development appears to be working. But in many other countries it is not.

At the bottom of the electricity totem pole are the approximately 1.7 billion people who currently live in Third World villages without lights, appliances, or other benefits of electricity.[4] Although in some countries such as China and the Philippines rural electrification is proceeding steadily, in most of Africa and parts of Latin America it has come to a virtual standstill as economic conditions have worsened in recent years. Furthermore, much of the rural electrification that has occurred has brought far fewer economic or social benefits than it could. Rural energy programs must be reorganized and revitalized if the limited funds available are to have a larger impact on agricultural productivity, the development of rural industry, and the improvement of village living conditions.

Most Third World electric utilities are government-owned monopolies, often having strong political connections and the power to commit large sums of money. However, Third World utilities are increasingly troubled institutions. Not only is their financial condition deteriorating, many have management problems caused by the pace of recent growth. Budget crises have forced substantial salary cuts and the loss of top engineers and managers. The first step in any effort to put Third World utility systems on a sustainable footing has to be increased attention to the basics of good management, including programs to hire, train, and keep qualified personnel.

Third World utilities also need to place greater emphasis on energy efficiency. From generation and distribution to the way that electricity is used, Third World power systems are among the world's least efficient, each year wasting billions of dollars worth of electricity. There is now convincing evidence that careful programs to invest in efficiency improvements can provide developing countries with electricity services at far lower cost than most new power sources being developed. To begin with, each utility should plan on devoting at least 10 to 15 percent of its capital budget to increased efficiency, chiefly in industry and commercial buildings.

Electricity planners have focused almost exclusively on large centralized power projects. And the long lead times and substantial interest costs involved in the construction of such plants have contributed to the financial problems of many utilities. Smaller plants, whether relying on hydropower, biomass, wind power, solar technologies, or the traditional fossil fuels, can be built more quickly in

7

8 response to consumption trends and without the need to tie up capital resources for five to ten years. Many of these energy sources are now more economical than large conventional power plants, and have the advantage of relying on domestic renewable resources rather than imported fuels.

The Third World's power institutions can be decentralized along with the generating plants. Although government monopolies now run all but a handful of the electric power systems in the developing world, efficiency could be improved and innovation spurred if a broader range of public and private institutions were involved. The institutional innovation that has been most widely tested is the rural electric cooperative, usually a member-owned distribution utility that brings power to villages. Co-ops can circumvent national utilities' frequent lack of interest in rural electrification and can serve as vehicles to introduce electricity and a range of other development tools to remote villages. In addition, private generating companies can be permitted to build and own small power plants, which encourages profit-motivated innovation and takes some of the financial and management burden off central power authorities.

There are many impediments to the reform of Third World electricity programs. Altering the status quo is always difficult, particularly if it involves a large and politically powerful sector of society. The sheer pace of ongoing construction also undermines attempts to change. With growing electricity needs still outstripping even the more optimistic construction schedules, policymakers are reluctant to do anything that would slow electrification programs from full-speed-ahead. But most developing countries simply cannot afford to neglect fundamental problems any longer. For better or worse, electric power is a large and crucial sector of most Third World economies, and if its future growth is misdirected or poorly managed, the development process itself could be jeopardized.

Developing Electric Power

Electricity development in the Third World began in the early part of this century, only a short time after similar efforts got under way in North America and Europe. These early systems, often owned by subsidiaries of U.S. or European power companies, mainly served large cities and grew quite slowly. But in the post-war rush to industrialize the Third World, development planners considered electric power to be an essential infrastructure investment and made it a top priority. The building of cities, the creation of modern industries, and even the expansion of raw materials exports require a growing power supply. Starting with very little electricity, officials of the newly nationalized utilities typically achieved growth rates as high as 20 to 30 percent per year.[5]

The support of international financial institutions has been an essential catalyst to Third World electricity development. Commercial banks and private companies were initially reluctant to participate in large projects sponsored by new utilities that had little technical or management experience. The high cost and substantial foreign exchange requirements of power projects make them among the most challenging of projects to finance. The World Bank and the various regional development banks have provided billions of dollars in loans to Third World electricity projects. These loans in turn have stimulated many more billions in commercial loans.[6]

World Bank support of electricity systems grew from $85 million annually in the mid-fifties to $271 million in the mid-sixties, $1,400 million in the early seventies, and $1,800 million in the early eighties. The Bank's support of electric power projects has leveled off in recent years and shrunk in proportional terms as lending expanded in other areas. Nonetheless, by mid-1984, the World Bank had committed a total of $24.4 billion to electric power development, about 16 percent of total Bank program expenditures. This makes electric power the third largest focus of World Bank lending after agriculture and transportation. However, since these latter areas include several different categories of projects, electric power development is actually the single largest purpose for which the World Bank lends money.[7]

Other development agencies and lending institutions have also generously supported Third World power projects. The Inter-American Development Bank loaned approximately $800 million for power projects in Latin America. in 1985, or about one quarter of its total lending. This is more than twice the amount lent for health, education, and urban development combined. The Asian Development Bank has loaned an average of $400 million annually for power projects since 1980. The much smaller African Development Bank loaned about $20 million annually for electric power in recent years, about one-fifth of total disbursements. Total development bank lending for power projects comes to about $3 billion annually.[8]

Many bilateral aid agencies in Europe and North America also support Third World power programs. Among the most prominent is the U.S. Agency for International Development which had major programs in Latin America and elsewhere until the aid program was redirected to meeting basic human needs in the seventies. French agencies also have helped develop electricity systems in West Africa.

Although foreign aid and low-interest loans stimulate electric power development in the Third World, most of the capital is generated internally or borrowed on international financial markets through commercial banks. International bankers usually consider established Third World utilities to be relatively creditworthy, in part because the national treasury effectively stands behind most power loans. No comprehensive figures are available, but total Third World investment in electric power since the fifties is probably over $500 billion (1985 dollars).[9]

Most developing countries are served by large nationally owned utility companies, many of which were taken over from foreign-owned private utilities in the fifties. Electricity systems dominated by private utilities, as in the United States and Japan, are now rare in the Third World.

> "No comprehensive figures are available,
> but total Third World investment in
> electric power since the fifties is probably
> over $500 billion."

Large countries and those with clearly distinct geographic regions often have more complex power systems. In Brazil, state-owned utilities supply much of the country's power, but the federally owned Centrais Eletricas Brasileiras supervises the generation and transmission of electricity. India has a power system whose complexity rivals that of the United States. It has a Central Electricity Authority that oversees national power planning, but the State Electricity Boards and Regional Electricity Boards actually run the country's power systems which are not yet fully integrated into a national grid. In the Philippines the National Power Authority owns virtually all of the power plants and transmission lines but the Manila Electric Company, originally set up as a subsidiary of the General Public Utilities Corporation of New York, distributes power in Metro Manila.[10]

The general trend is toward greater centralization and governmental control of electric power systems. Both commercial banks and government-supported lending agencies prefer to deal with a strong central authority that has government financial backing but is outside the day-to-day political process. Some early projects flopped because they were inadequately managed or because the utility did not have the authority to take actions essential to the success of its programs (such as raising electricity prices). Once a single power authority is established, it is easier to ensure that highly trained technocrats using the latest management techniques are running the program.

World Bank files document a consistent push for greater centralization and consolidation of authority whenever questions of the structure of a power system arise. Over the years, the World Bank has gradually become stricter in the institutional preconditions it sets for power loans, even occasionally suggesting who should be appointed to key positions in the national power authority. This has naturally angered political leaders who believe that this kind of involvement sometimes impinges on national sovereignty.[11]

Electric utilities are without doubt among the strongest institutions in developing countries, thanks both to the magnitude of their financial resources and to their partial autonomy within the political system.

Utility officials have a generally higher level of education and show greater understanding of basic principles of management than in most Third World ministries (though often not as good as in private industry).[12] Electric power systems are extremely complex technologies in which breakdowns are a constant threat. That many low-income developing countries have been able in a short period to build up sizable electric power systems is a substantial achievement.

By the early eighties, Third World nations were using six times as much electric power as they had two decades earlier. But compared to industrial countries, electricity plays a relatively small role in Third World economies. Wood and crop wastes still provide developing countries with at least 12 times as much energy as electricity does. As of 1982, annual per capita electricity use ranged from a high of 1,402 kilowatt-hours per person in Argentina and 1,192 kilowatt-hours per person in South Korea to 36 kilowatt-hours per person in Bangladesh and 23 kilowatt-hours per person in Nepal, enough to run a 30-watt bulb for a month. (See Table 1.) These figures contrast with electricity consumption of about 4,500 kilowatt-hours per person in Europe and 9,600 kilowatt-hours per person in the United States.[13]

Wealthier countries usually consume more electricity but even in countries with relatively high per capita use, a large share of the population may be completely without power. Over half of the electricity produced in most developing countries goes to industry, a far larger fraction than in the industrial world. And developing countries tend to use most of the electricity in just a handful of power-intensive industries, often those that produce goods for export. In Mexico, 55 percent of the electricity is used by industry, and in South Korea, 68 percent is.[14] A growing amount of electricity is also used in the modern office buildings and hotels that dot the skylines of many Third World capitals.

Household use of electricity is quite limited in most developing countries. Many Third World consumers cannot afford to purchase, let alone operate, appliances that are heavy users of electricity. Even in cities, only the wealthiest of families can pay for refrigerators, electric

Table 1: Electricity Use in Selected Developing Countries, 1982

Country	Per Capita Income (U.S. $ per year)	Electricity Use (million kilowatt-hours)	Per Capita Electricity Use (kilowatt-hours)
Argentina	2,520	39,804	1,402
Brazil	2,240	151,721	1,197
South Korea	1,660	47,197	1,192
Mexico	2,737	80,589	1,103
Costa Rica	1,806	2,500	1,041
Zimbabwe	849	7,614	1,015
Colombia	1,300	22,564	837
Philippines	746	20,560	405
Nicaragua	802	1,153	400
Thailand	764	17,687	365
China	307	327,678	325
Bolivia	932	1,703	290
India	247	138,677	197
Zaire	203	4,392	143
Kenya	420	1,998	110
Senegal	541	633	105
Indonesia	524	12,722	83
Niger	338	350	60
Bangladesh	132	3,305	36
Nepal	136	356	23

Source: World Bank, "1982 Power/Energy Data Sheets for 104 Developing Countries."

stoves, air conditioners, and the other appliances that many consumers in the industrial world now take for granted. Lightbulbs, television sets and fans place the heaviest demand on power in a Third World home. Residents of Manila or São Paulo typically pay as much for electricity as do their counterparts in Washington or Paris.

But because their incomes are far smaller, their electric bills consume a proportionately larger share of their incomes.

Electricity consumption in most developing countries is so low and the potential future uses so great that electricity use continues to expand even when the economy does not. The World Bank estimates that Third World electricity use will increase 7 percent annually in the next decade, and many countries forecast more rapid growth.[15]

Similar demand projections led to the first big wave of power plant construction in the Third World in the sixties. Many oil-fired plants were built because they were relatively inexpensive compared to hydropower facilities. Since imported oil then cost only $2 to $3 per barrel, few planners had any qualms about relying on it. Starting with relatively insignificant power systems, most Third World planners estimated a need to expand electricity systems at well over 10 percent annually. This massive construction effort was in the process of gearing up when oil prices soared in the early seventies.

Looking for alternatives to oil-fired plants, Third World planners turned with renewed interest to hydropower construction. Many developing countries have excellent hydropower potential, which is usually associated with mountainous terrain and abundant rain or snow. If managed properly, hydropower is a renewable energy source immune from future fuel price increases. Whereas North America and Europe had developed 59 percent and 36 percent of their hydropower potential respectively by 1980, Asia had harnessed just 9 percent, Latin America 8 percent, and Africa 5 percent.[16]

Third World hydropower development involves some of the largest and most expensive civil works projects in history, including the 10,000-megawatt Guri project in Venezuela and the 12,600-megawatt Itaipu plant on the Brazil-Paraguay border, both of which are still under construction. The latter project will produce as much power as 13 nuclear plants. Between 1978 and 1983, Mexico's hydropower capacity rose 43 percent, Brazil's 55 percent, and Argentina's 58 percent.[17]

> **"Electricity consumption in most developing countries is so low and the potential future uses so great that electricity use continues to expand even when the economy does not."**

The World Bank projects that between 1980 and 1990, Third World hydro capacity (excluding China's) will more than double—rising from 100,000 megawatts to 201,000 megawatts. China alone has 17,000 megawatts of large hydro projects under construction that will bring the country's total to 41,000 megawatts. The proposed Three Gorges project on the Yangtze River would provide additional capacity of about 13,000 megawatts.[18]

15

The environmental and human costs of some of these hydropower projects is substantial. Construction of new dams in developing countries has displaced millions of people, flooded agricultural land, and trapped in reservoirs river silt that once fertilized flood plains. The visions of power planners are now often in direct conflict with the needs of indigenous people. The Kariba Dam built in Zimbabwe in the sixties uprooted 56,000 people, many of whom never found suitable homes, farmland, or clean drinking water. Many died of dysentery and those relocated still rely on food imports. The Chico River project in the Philippines displaced thousands of native people and led to an armed struggle with government authorities.[19]

The Three Gorges project in China requires planners to consider unprecedented trade-offs. The proposed facility could increase the country's power capacity by 15 percent, reduce flooding downstream, and provide additional irrigation for the rich Sichuan Plain. But Three Gorges would also flood thousands of hectares of land upstream of the dam and displace between 300,000 and one million people depending on the height of the dam. This part of China is one of the most agriculturally productive and densely populated areas of the world. Some provincial authorities vehemently oppose the project.[20]

Third World governments also wish to increase coal's contribution to electricity generation. The World Bank projects an expansion of coal-fired generating capacity (excluding China) from 35,000 megawatts in 1980 to 92,000 megawatts by the year 2000. Most of this development is concentrated in the relatively few Third World countries with substantial coal reserves, such as India and Colombia. China, which has coal reserves equal to those of the Soviet Union and the United States combined, has plans to boost its coal-generating capacity of about 50,000 megawatts to over 140,000 megawatts by the year 2000. China has already converted most of its 12,000 megawatts of oil-fired generating capacity to coal.[21]

Even more than hydropower, coal development will involve large environmental trade-offs. Northern Chinese cities are already heavily polluted by coal burning, and evidence of acid rain damage is beginning to appear in several Third World countries. The high cost and complexity of pollution control technologies have been obstacles to their widespread use in many developing countries. In addition, coal combustion is a major contributor to atmospheric carbon dioxide levels which many scientists believe are increasing so rapidly as to severely alter the earth's climate within 30 to 50 years.[22]

Third World power systems are still heavily dependent on fossil fuels, though hydropower is the predominant electricity source in many countries. (See Table 2.) Most of the largest developing countries, with the important exceptions of Brazil and Colombia, obtain at least 60 percent of their power output from fossil fuels and many are still heavily dependent on oil. Overall, figures for 1980 show hydropower supplying 38 percent of Third World electricity, coal supplying 30 percent, and oil 26 percent.[23]

Nuclear power, once widely promoted in the Third World, has lost ground in many nations. For developing countries, nuclear power presents the problem of not only being complicated and expensive but requiring more foreign exchange than do most power investments. Major projects in Argentina, Brazil, and the Philippines have run into technical and financial problems; political disputes drastically slowed nuclear programs in Iran, Iraq, and Pakistan.[24]

Only in the rapidly industrializing Far East has nuclear power contributed significantly to power supplies. South Korea has 2,865 megawatts of nuclear capacity or 18 percent of its total, and Taiwan has 5,146 megawatts. (See Table 3.) Both nations plan to continue

Table 2: Electricity Generating Capacity in Selected Countries, 1982

Country	Total Capacity	Fossil Fuel	Share Hydro	Geo-thermal	Nuclear
	(megawatts)		(percent)		
China	72,360	68	32	0	0
Brazil	38,904	15	85	0	0
India	38,808	64	34	0	0
Mexico	22,574	68	31	1	0
Argentina	13,460	63	35	0	3
South Korea	11,597	79	10	0	11
Philippines	5,054	64	25	11	0
Thailand	4,694	71	29	0	0
Colombia	4,660	36	64	0	0
Indonesia	3,513	83	16	1	0
Zaire	1,716	3	97	0	0
Zimbabwe	1,192	41	59	0	0
Bangladesh	990	92	8	0	0
Costa Rica	657	30	70	0	0
Kenya	574	33	62	5	0
Bolivia	508	44	56	0	0
Nicaragua	400	65	26	9	0
Senegal	165	100	0	0	0
Nepal	162	22	78	0	0
Niger	100	100	0	0	0

Source: World Bank, "1982 Power/Energy Data Sheets for 104 Developing Countries."

their nuclear construction programs. But no other developing country produces significant nuclear power. The Third World as a whole gets less than 4 percent of its electricity from nuclear power, and that figure is unlikely to increase significantly by the year 2000.[25]

Table 3: Nuclear Power in Selected Developing Countries, 1986

Country	Nuclear Capacity	Share of Total Capacity
	(megawatts)	(percent)
Taiwan	5,146	48
South Korea	2,865	18
India	1,330	3
Argentina	1,005	7
Brazil	657	1
Pakistan	137	3

Source: *Nucleonics Week*, January 30, 1986.

For several years, China has been on the verge of a major nuclear construction program involving as much as 10,000 megawatts by the year 2000, part of a plan to quadruple electricity supplies by the end of the century. Although China has developed its own nuclear technology, joint ventures with foreign firms were planned to accelerate nuclear development. Most of the industrial world's active nuclear vendors and sponsoring government officials have traveled to China to pursue one of the few remaining international markets. Although China's leaders express a continuing commitment to the nuclear program, they decided in 1986 to delay construction of all but two plants until the nineties. China's leadership appears to have decided that it cannot afford simultaneous large hydro and nuclear programs and that given the lower foreign exchange cost of hydro development, it deserves priority.[26]

Some developing countries are attempting to build strength through diversity. One of the most successful efforts at electricity diversification is that of the Philippines. The Philippines has reduced the oil share of electricity generation from 78 percent in 1978 to 46 percent in 1984 through the aggressive development of hydro and geothermal resources. By 1984, the Philippines had 1,654 megawatts of hydropower capacity in place and 894 megawatts of geothermal capacity out of a total of 5,196 megawatts. An additional 604 megawatts of hydro dams, 110 megawatts of geothermal, and 110 megawatts of coal are under development. Several small power plants that burn wood and coconut shells are operating or under construction. However, the investment in all of these energy sources is dwarfed by the $2 billion the Philippines has spent on a single nuclear plant that will probably never operate because of safety concerns.[27]

The cost of building additional generating capacity in the Third World has increased dramatically since the mid-seventies. This is due in part to the fact that hydropower and coal plants are inherently more expensive than oil plants, and also because developers have had to turn to more costly hydro projects as the most attractive sites are developed first. In the sixties, hydro projects costing more than $800 per kilowatt were considered economically questionable, but now many developing countries are building facilities that cost $2,000 to $3,000 per kilowatt.[28]

Construction cost increases have coincided with a period of high interest rates and general economic deterioration in many developing countries. Typically, about one third of a power project's cost requires foreign exchange payments to outside companies, a burden made all the worse by large interest payments. Many Third World utilities must also pay large bills for imported oil. Since most Third World utilities are wards of the state, the national treasury stands behind power loans, and utility debts become entwined with the national debt.[29]

Although no precise figures are available, total Third World utility debt can be estimated at over $180 billion, or one-fifth of the Third World's accumulated debt of over $900 billion. A 1985 report to the U.S. Agency for International Development on the situation in Central America concluded that, "the region's financial crisis, and its constraints on future development, is significantly the result of huge public power investments in nearly all of the countries." In Costa Rica, 18 percent of the country's foreign debt is attributed to the power sector and in Honduras 33 percent is.[30]

The World Bank projects that developing countries will have to invest $60 billion in electric power each year in order to keep up with a demand growth rate of 7 percent.[31] This is more money than the Third World receives in development assistance funds annually. These investment goals are for the most part unattainable, and many utilities are simply not keeping up with demand. Combined pressures of underfinancing and overuse result in inadequate generating capacity and periodic planned or emergency blackouts as power systems become overloaded.

Financial pressures often cause utilities to cut back on maintenance work which causes a gradual deterioration of reliability over time. Transmission and distribution systems often suffer the most. They receive less attention than power plants, and their upkeep is usually the first item cut from the budget. In many countries over 30 percent of the power transmitted is lost because of inadequate maintenance, undersized lines, and illegal tapping of power lines. Wide voltage fluctuations that damage appliances are common in Third World power systems, leading industries to compensate by purchasing oversized electric motors that use additional electricity. This increase in demand further strains power systems and adds to the financial burdens of power authorities.

Pakistan faces some of the worst problems: Losses on the national power system now average 38 percent and blackouts occur daily. Many companies must maintain backup generators in order to keep their plants running. A shortage of electricity is now considered to be one of the most serious roadblocks to Pakistan's economic development. China's leaders also project that its electricity supplies (growing 6 percent annually) will not keep up with needs. Already in Sichuan Province power is cut three days a week to ensure adequate power for the other four.[32]

The problems of many electric utilities are getting worse. The World Bank has projected a $32 billion financial gap between utility needs and financial capabilities in Latin America and the Caribbean during the next five years. In some cases, the cost of servicing the utility's

> "The World Bank projects that developing countries will have to invest $60 billion in electric power each year in order to keep up with a demand growth rate of 7 percent."

debt now exceeds total operating revenues; construction is continually adding to the debt load. Chile is fairly typical with a planned construction budget of $1.4 billion for 1984-1988 and scheduled interest payments of $1 billion, but projected operating revenues of just $1.1 billion.[33]

Increasingly, government leaders are projecting financial shortfalls and forcing utilities to curtail construction spending. Recent budget negotiations in Brazil and India forced power authorities to cut planned spending by half. A study of Central American utilities concludes that, "Due to a combination of existing debt obligations and internal organizational problems, the large national utilities could have difficulty financing planned power investments, beginning immediately." China has so far avoided large debts but it now faces a substantial shortage of electricity that will constrain its economic growth. Planners project a need to increase capacity fourfold between 1980 and 2000, which will be a large financial burden.[34]

But the problem is not simply one of money. Many Third World power systems have also outgrown their own internal management capabilities and are increasingly short of technical expertise as well. One African country mistakenly spent $2 billion on a power plant and transmission line that it will not need for decades. Some utilities, having relied for years on expatriate labor and temporary consultants, are now short of people skilled in specialized fields. Many of the best-trained people have left for better-paying jobs in the private sector or overseas. The complex, demanding power systems in developing countries will fall apart unless they can attract and keep motivated, trained individuals.[35]

While some utilities are efficiently managed, politically neutral organizations, others are managerial disasters plagued by petty politics. Many have become overly bureaucratic, with the lack of innovation and slowness to respond that is typical of institutions not subject to market forces. Some Third World utilities are described as employment agencies that are forced by governments to overhire in order to create jobs. Corruption is not uncommon and is magnified by the quantities of money involved. The multimillion-dollar payoffs that allegedly passed from the U.S.-based Westinghouse Corporation to the Marcos government in the Philippines in order to secure a contract for the Bataan nuclear plant is an extreme case, but many experts believe it reflects a more widespread problem.[36]

Selectively trimming construction budgets, strengthening maintenance procedures, raising salaries, and introducing better management techniques are essential to any serious program of improvements of Third World power systems. Internal changes and external pressure from lending agencies are causing many utilities to carry out these improvements. However, to meet the ever-growing demand for energy, planners will need to make even more fundamental changes to power development and distribution. Improved efficiency, increased reliance on decentralized technologies, and wide-ranging institutional reforms must be incorporated into efforts to improve Third World electricity systems.

The Efficient Use of Electricity

A 1986 study of Third World utilities notes that, "The atmosphere in LDC [Less Developed Country] utility boardrooms is frequently very similar to their counterparts in the U.S., in the love of ever larger capital construction programs, and the unshakeable belief that electric demand will continue to grow at 10 percent per year forever."[37] Managing a utility is often equated with building power plants and transmission lines, and indeed this task is so overwhelming in many countries that the demand for electricity is unquestioned.

Industrial countries are now facing a rather different situation. Since the mid-seventies, demand for energy in general and electricity in particular has grown more slowly, confounding the forecasters. In part, this is because of slower economic growth and a shift to less energy-intensive industries. But the main reason is improvements in electric efficiency caused by higher prices. In the United States, for example, electricity consumption per dollar of gross national product increased quite steadily until the mid-seventies, peaked in 1976, and has fallen slightly in all but one year since. Growth in the consumption of electricity has averaged just 2 to 3 percent annually in most industrial countries since 1980, leaving many nations with excess generating capacity.[38]

The long-held belief that electricity growth has a fixed relationship with economic growth is no longer considered valid, and analysts must now incorporate efficiency projections into forecasts. Varying efficiency levels can over a period of a decade make a large difference in the amount of required additional generating capacity. Forecasting is complicated by the fact that it is difficult to know how quickly efficient technologies will advance, and how rapidly consumers will adopt the new devices. Nonetheless, many sophisticated end-use based forecasting models now incorporate efficiency projections into power planning in industrial countries.

Even though the importance of electric efficiency has been accepted in industrial countries, its potential has been neglected in the Third World. This is undoubtedly a result of the explosive electricity growth still occurring in most developing countries which appears to dwarf any potential contribution from greater efficiency. However, this neglect makes little sense. When the inventory of electricity-using equipment is growing quickly, there is greater potential to improve the overall level of efficiency. If energy efficiency is economically attractive when electricity use is relatively stable, it is more attractive when use is rising since it can displace some of the most expensive planned generating capacity. (In industrial countries, on the other hand, electricity conservation must sometimes be justified by its potential to economically displace existing capacity.) Looked at another way, improved energy efficiency raises the productivity of new power plants.

Research and development have increased the efficiency of many technologies in recent years. In 1985, a U.S. study found that in the past decade the efficiency of new refrigerators increased 52 percent and that of room air-conditioners 76 percent. Even greater improvements have occurred in Japan, because Japanese companies have marketed the new appliances much more enthusiastically than their American counterparts. New Japanese and European fluorescent lights use one third as much power as systems now in use. Electric motors, the largest user of electricity in most industries, have been improved in a number of ways. In some cases, adjustable speed drives can decrease the power use of electric motors by 20 to 30 percent. Newly developed aluminum smelters reduce the electricity required in this power-intensive process by 24 percent, and aluminum recycling can lessen power needs by 95 percent.[39]

A 1980 World Bank study, *Energy in the Developing Countries*, included an extensive analysis of "The Demand for Energy and its Management." Among the study's conclusions was the observation that, "now that energy is no longer cheap or abundant, energy efficiency must be considered as a principal element in economic planning and energy demand management must take its place with other forms of economic management."[40] Other development agencies and many Third World energy ministries have come to similar conclusions, although most of the emphasis is placed on saving oil.

A gap has opened between awareness of the potential to save electricity and the development of effective programs. This gap is undoubtedly caused in part by a shortage of data on the uses of electricity. Without such information it is difficult to justify energy-efficiency programs, let alone carry them out. But, efficiency programs are often "institutional orphans." Agencies that might logically have some interest in such programs include the ministries of energy, industry, and housing, as well as the power authority and public and private corporations. Most agencies would rather not take on additional responsibilities, and often new tasks with crosscutting jurisdictions do not get done at all.

Energy analysts are developing data on the way electricity is used in the Third World which hint at the potential for energy-efficiency programs. Since the proportion of electricity used in industry typically ranges from 40 to 70 percent, it deserves the closest scrutiny. A study in Brazil found that seven industries, producing 13 percent of the country's industrial goods, use half of industrial electricity.[41] Major aluminum producers such as Brazil, Ghana, Surinam, and Venezuela devote a large share of electricity to the energy-intensive smelting process. In most other industries the largest consumer of electricity is the electric motor, used to drive a range of mechanical processes.

The commercial areas of many Third World cities are now filled with highrise buildings and hotels that use large quantities of electricity. Although only a small fraction of the people use such facilities, many Third World capitals consume as much power as all of the country's villages combined. Since the buildings are often based on foreign designs intended for a temperate climate, they are inappropriate to the year-round warmth and humidity found in tropical countries.

Agriculture also makes a significant claim on the electricity systems of some developing countries. Much of this is for water pumping for irrigation in semi-arid regions. In India, for example, 18.6 percent of the country's electricity is estimated to go to irrigation pumping.[42] In many rural areas, over half of the electricity is used for pumping.

In most cases, it takes more electricity to perform a given task in developing countries than in the industrial world. There are several reasons for this. Third World nations generally use equipment that was designed years earlier in another country, a clear disadvantage when the efficiency of new appliances is advancing. In some industries the manufacturing process itself is outmoded and much more energy-intensive than newer processes. In addition, inadequate equipment maintenance can lower efficiency. Finally, many of the devices used are inappropriate for the designated task; for example, oversized electric motors are deployed to compensate for voltage fluctuations. Improving the operating reliability of power systems may have a double advantage if it encourages the use of more efficient motors.[43]

Third World countries would realize substantial economic benefits if electricity was used more efficiently, because it would lower production costs and the price of goods. Exports would therefore be more competitive on world markets. Increasingly, energy-inefficient products and processes will put developing countries at a major disadvantage.

Meeting projected growth in the demand for electricity services will be virtually impossible without substantial efficiency improvements. Without efficiency gains, Brazilian planners project that the country's generating capacity will have to grow 150 percent in the next 15 years at a cost of over $130 billion. Of the 66,000 megawatts of additional projected power needs, only about 34,000 megawatts is currently planned. If some portion of the remaining 32,000 megawatts of forecasted power needs was met by improved efficiency, capital requirements could be reduced and industrial productivity raised. A 1985 study concludes that projected electricity consumption just in selected major power uses in Brazil could be cut 30 percent by the year 2000 if a range of cost-effective efficiency measures were adopted. (See Table 4.) The $10 billion needed to implement these measures would eliminate the need for 22,000 megawatts of generating capacity that would cost an estimated $44 billion.[44]

The cornerstone of any program to improve efficiency is a pricing system that reflects the true cost of providing power. For most goods and services, the interaction of supply and demand determine price, but electricity prices are usually set by government policy, with the goal of balancing the desire of consumers for lower prices and the need to keep the utility financially healthy. Electricity pricing often becomes a political tool, with prices lowered in advance of elections in order to ensure support for incumbent politicians. Some subsidizing of electricity prices is inevitable and beneficial. Few villagers would ever get electricity if in the first few years they were forced to pay the full cost of providing power. The problem comes when low, subsidized electricity prices continue indefinitely. Naturally, consumers will use greater quantities of something that costs them relatively

"The cornerstone of any program to
improve efficiency is a pricing system that
reflects the true cost of providing power."

Table 4: Potential Electricity Savings in Selected End-Uses[1] in
Brazil by the Year 2000

Electricity Use	Current Forecast	Savings Potential	Savings Potential
	(million kilowatt-hours)		(percent)
Industrial Motors	177.3	35.5	20
Domestic Refrigerators	28.3	14.8	52
Domestic Lighting	17.7	8.8	50
Commercial Motors	29.7	5.9	20
Commercial Lighting	25.8	15.5	60
Street Lighting	17.9	7.2	40
Total	296.7	87.7	30

[1]These account for about two-thirds of total electricity use in Brazil. These projected savings include only efficiency improvements that are economically justifiable based on currently available technologies.

Source: Howard Geller, "Electricity Conservation in Brazil."

little; higher consumption in turn leads to the need to expand the power supply which raises costs.

Economists use the term "marginal cost pricing" to describe a system that charges the consumer just what the utility must pay to deliver additional electricity. This system should provide the consumer with

28

incentives to use electricity efficiently and result in the "right" level of power use and plant construction. Since the seventies, the World Bank, International Monetary Fund, and other lending agencies have strongly advocated marginal cost pricing of electricity, though they view it more as a way to ensure the financial stability of utilities than as a way of encouraging energy efficiency. As described by World Bank economists Mohan Munasinghe and Jeremy Warford, "National economic resources must be allocated efficiently, not only among different sectors of the economy, but also within the electric power sector. This implies that prices that reflect cost must be used to indicate to electricity consumers the true economic cost of supplying their specific needs, so that supply and demand can be matched efficiently."[45]

Many countries have adopted marginal cost pricing of electricity in recent years. Prices have been raised, and many countries now have "time-of-use" rates for industry in order to discourage use of power at times of peak demand. But complexities and disputes have slowed the move to price reform. Often the cost of providing power varies greatly even within a country. In addition, prices are sometimes subsidized in order to make particular industries competitive or to make electricity competitive with subsidized diesel fuel. Additional reforms are needed, but pricing will inevitably continue to be a compromise between the goals of planners and the needs and demands of consumers.[46]

Misdirected pricing policies are not the only obstacles to the more efficient use of electricity. In many developing countries the government guarantees the purchase of manufactured goods on a cost-plus basis. An improvement in efficiency and the resulting cost-saving do not increase the company's profit margin. Even where more efficient technologies are clearly economical based on current electricity prices, studies show that consumers often purchase a less efficient appliance. People who have virtually no disposable income cannot afford to buy a slightly more expensive but efficient device even if the investment would pay for itself in less than three years.

There is a broad range of programs that can help overcome barriers to improved energy efficiency. The simplest are educational campaigns to encourage industries and consumers to make informed decisions about power usage. Campaigns can promote efficiency labels on appliances, free energy audits for factories or homes, and even simple reminders to turn out lights and maintain appliances. Second, governments can initiate mandatory efficiency standards for appliances, homes, and industrial equipment. Third, governments or utilities can directly assist in the financing of efficiency investments, either using low-interest loans or outright grants. Many countries have tried variations on these approaches and developing nations no longer need look to the industrial world to find examples of programs that work.

South Korea has one of the highest electricity consumption growth rates in the world and has severely limited domestic energy reserves. Not surprisingly, the Korean government has initiated one of the most comprehensive efficiency programs of any nation. In 1980 it enacted a "Law Governing Rationalization in the Use of Energy." Mandatory building codes specify the energy efficiency of all new buildings; efficiency labels are required on all new household appliances; a national energy conservation center was established to provide technical support and training. In addition, the government has introduced financial incentives for energy conservation. These include income tax deductions, reduced tariffs on energy-saving equipment, a one-year depreciation allowance for energy-saving investments, and subsidized loan programs for conservation investments in industrial plants and buildings.[47]

Although South Korea has an increasingly market-oriented economy, it retains the Confucian tradition of central control. If government efficiency standards are deemed to be in the national interest, they are accepted. Confronted with the prospect of power shortages, the government has simply curtailed some power uses. For example, it has banned the use of air conditioners except for a 40-day period in the summer and prohibited the use of elevators in the first three floors of buildings. Saving electricity is considered a national priority, because the nation's ambitious economic plans are to some extent dependent on maintaining competitive electricity prices. As South Korea develops economically, the competitive advantage of low wages will diminish, and rising energy costs could price their products out of some markets.

Brazil, the Philippines, and Singapore are among the other developing countries that have made energy efficiency a national priority and have established substantial programs. Among Brazil's programs are an energy labeling program for refrigerators, and an electricity price system that penalizes heavy consumers. In the Philippines, the National Bureau of Energy Utilization provides free energy audits to large factories and commercial buildings, and efficiency labels are required on air conditoners. In addition, a U.S.-supported pilot program assists in the financing of energy conservation. Singapore introduced a financial incentives program for energy-efficient buildings in 1979.[48]

One of the problems many countries face is how to encourage the manufacture of more efficient equipment. Companies often perceive that consumers care little about efficiency and so they neglect potential innovations. In some developing countries, companies manufacture efficient equipment for export while selling less efficient devices domestically. Efficiency standards are probably needed in most countries and can be met by domestic research or by obtaining foreign technologies through manufacturing licenses. Brazil now has a government program to support the development of more efficient refrigerators, lighting systems, heat pumps, and air conditioners.[49] Internationally certified appliance-testing programs would be a big help. With the international marketing of appliances growing rapidly and several countries already having incompatible standards, future innovations may be hampered without some standardization.

It is too early to assess the final impact of energy-efficiency programs. Data collection is often inadequate and sorting out the relative importance of different measures is difficult. Nonetheless, there is every reason to believe that many of these programs will yield substantial benefits. However, more effort could be made to target technologies that make up the bulk of power use in particular countries. By concentrating, for example, on lighting, water pumps, and electric motors, large amounts of power could be saved at relatively small cost.

One of the most difficult problems is how to fit electricity-efficiency programs within existing government bureaucracies. The ministries of industry, housing, and energy can each claim jurisdiction as can the power authority and various state or local agencies. Resolving overlapping claims and designating one agency as the lead for electricity conservation is essential if efficiency efforts are to be coordinated. Brazil has taken steps in this direction by starting a national program for electricity conservation in 1985, and India has developed an interministerial strategy for electricity and fuel saving.[50]

Striking a balance between mandated standards and market-oriented financial incentives is also important. Standards are the most effective way of ensuring rapid achievement of certain minimal efficiency levels, but they will not encourage ongoing research to drive electricity requirements ever lower. And because energy markets do not function well in many developing countries, the use of market incentives may not be effective without wider economic reforms.

Efficiency programs could benefit from the technical competence and managerial expertise of national power authorities. The utilities can also ensure that efficiency will take an important place in the planning process, alongside the construction of power plants and transmission lines. Since efficiency investments can in effect provide additional services in the same way that power plants do, utilities should establish a policy to invest in efficiency whenever it is cost-competitive with new generating options under consideration. If efficiency and production programs are kept entirely separate, a utility could be building a power plant that costs $2,000 per kilowatt, while the energy ministry is passing up efficiency investments that cost $500 per kilowatt.

Utilities can also help to finance and organize efficiency programs, a concept that has been successfully pioneered by several utilities in the United States. The Pacific Gas and Electric Company (PG&E) in northern California, for example, has a peak demand of 13,000 megawatts, more than that of all but five Third World nations. PG&E sponsors extensive energy audits, offers zero-interest loans for home weatherization, and has rebate incentives for the purchase of energy-saving devices. PG&E economists calculate that the $80 million spent in 1983 resulted in a savings of 240 megawatts. The savings cost of $350 per kilowatt contrasts with the $2,500 per kilowatt cost of PG&E's recently completed Diablo Canyon nuclear plant.[51]

The Austin Municipal Utility in Texas has taken conservation a step further by considering its efficiency programs to be a "conservation power plant." The programs are assessed and organized just as carefully as if they were building new capacity; the conservation programs are expected to yield a definite amount of power at a predetermined maximum cost. The 553-megawatt Austin program consists of utility payments to customers for investments in weatherization and the replacement of inefficient appliances. The customers are charged for the investment through their electric bills over a several year period, but because power needs are reduced, they pay less in total than they would have before the improvement was made. Everyone benefits because the utility has avoided building a coal-fired plant that would have cost $600 million more than the conservation efforts.[52]

The São Paulo utility, Brazil's largest, is one of the few Third World utilities to adopt a modest package of energy-efficiency programs. It is collecting data on the end-uses of electricity and is demonstrating and monitoring the use of energy-efficient electrical equipment. The Public Utilities Board of Singapore provides energy audits for some of its customers. In the Indian state of Karnataka, the state electricity board has set up an "electricity queue" which gives efficient companies priority in gaining access to the state's limited power supplies.

To date there appear to be no Third World utilities that are actively investing in electricity efficiency. This is unfortunate. Third World consumers are chronically starved for capital and cannot afford the initial efficiency investment even though the payback period is short. National utilities have access to relatively low-interest funds; re-channeling 10 to 15 percent of their construction budgets to efficiency programs could yield enormous returns. Consumers would be better served if electric utilities were gradually converted into energy service companies that would provide efficiency as well as new power supplies using the same economic criteria.

33

New Approaches to Rural Electrification

Third World electricity programs are largely oriented to providing power for cities and industries. Although 70 percent of the Third World's population lives in rural areas, political power flows to the cities, and electric power flows with it. Overall, fewer than one third of the people who live in the rural Third World now have access to electricity, leaving over 1.7 billion who do not. Nonetheless, rural electrification has become integral to the development process and each year millions more people are provided with electricity. Unfortunately, many rural electrification programs are underfunded, misdirected, or poorly managed. In many cases rural electrification is promoted as an end in itself, rather than as a means of reaching more basic goals.

When Costa Rica applied for a rural electrification loan in the fifties, the World Bank responded, "It is uneconomical to bring electricity to the people. Let the people move to the electricity."[53] However, this explicit urban bias lost favor in the sixties, and as electricity grids connected cities and industries, utilities began to run short distribution lines into villages. Many power authorities were forced by their national governments to electrify rural areas, often with external financial assistance. Some countries even established independent rural electrification agencies.

Politicians have found that electrification programs are popular and help their careers. Electrification is also seen as a way of unifying countries with different ethnic communities, thereby consolidating political power. Bangladesh's constitution includes a guarantee that all villages will eventually get electricity. However, as of 1981, only 3 percent of the country's villages had reached the constitutional goal. In most developing countries there is a gap between rhetorical support for rural electrification and the minimal funds devoted to its achievement. Overall, less than 10 percent of the Third World's electricity investments go to rural areas, and in many countries that investment is less than 5 percent.[54]

Comprehensive statistics on the extent of rural electrification are not generally available, though figures on the share of the total population with electricity are. (See Table 5.) From these statistics it can be surmised that the highest rates of rural electrification are in Taiwan and South Korea, and the lowest rates are in the African countries of Kenya and Niger and in Bangladesh and Nepal on the Indian subcontinent. Overall, just 5 percent of Africa's rural people have electricity. In Latin America, populations are more concentrated in cities where they generally have power, but over two-thirds of the rural population lacks electricity. However, Costa Rica is an exception. Its ambitious rural electrification program, begun in the early sixties, has reached 45 percent of the people.[55]

Several Asian countries have made great strides in rural electrification in the past 15 years. Virtually all of Taiwan has electricity, and the rural power systems are among the best managed in the world, a reflection of the general affluence and education of the Taiwanese. In India, 350,000 villages out of more than 600,000 (over 55 percent) are now connected to an electricity grid. The share is much higher in some parts of India. All of the villages of several states, including Kerala, Punjab, Haryana, and Tamil Nadu now have electricity. (However, even in "electrified" villages, often half or more of the houses lack power.) China has electrified 500,000 of its 710,000 villages, most of them in the past two decades. The amount of electricity available to rural areas has increased tenfold since 1965. Elsewhere in

Table 5: Extent of Electrification in Selected Developing Countries

Country	Share of Population in Rural Areas (1985)	Share of Total Population with Electricity (1982)
	(percent)	(percent)
Taiwan	29	99
Singapore	0	99
South Korea	43	95
Mexico	30	81
Costa Rica	52	80
China	79	60
Brazil	32	56
Colombia	33	54
Philippines	63	52
Senegal	58	36
Indonesia	78	16
India	77	14
Kenya	84	6
Nepal	94	5
Bangladesh	85	4
Niger	84	3

Sources: Population Reference Bureau, "1985 World Population Data Sheet"; World Bank, "1982 Power/Energy Data Sheets for 104 Developing Countries."

Asia, progress is slower. Less than 5 percent of Nepal's rural population has electricity, and in Bangladesh, less than 2,000 of 65,000 villages are electrified.[56]

In general, the more prosperous an area, the more likely it is to have electricity since people are more able to afford the hook-up charges and the cost of electricity. One way of calculating the economic feasibility of electrifying a region is to calculate its income density; that is, to multiply the number of people per acre by the per capita income. On the rich, well-watered agricultural plains of the Far East, rural electrification is easier to justify than in Africa's arid Sahel where both the population density and income levels are far lower. Isolated mountain communities, such as those in Nepal or Bolivia, are also hard to reach with electricity because population density is low and the rugged terrain makes the installation of power lines more difficult.

Electricity can bring sweeping changes to the lives of rural people. It often opens villages to the outside world and gives people the idea that things can change. Surveys show that many people look back on the arrival of electricity as a turning point in their lives. Electric lights are usually the first appliances purchased, a big improvement over gas or kerosene lamps. Electric lighting allows school children to read in the evening and extends the work day into the evening hours. Electric irons are also popular in many communities, as are radios, television sets, and electric fans.[57]

Most rural people do not have appliances such as refrigerators, washing machines, or electric stoves that are considered necessities in many industrial countries. Rural consumers cannot afford to purchase large amounts of electricity. The subsidized "lifeline" rates, available for limited electricity users, often do not apply if "luxury" appliances are purchased or if consumption rises above a minimal level of 30 to 50 kilowatt-hours per month—one-tenth the typical level of use in the United States. An even more important barrier is the cost of the appliances themselves which is usually not subsidized and is far beyond the means of most rural families. Even in Costa Rica and Colombia where villages are relatively prosperous, lights, irons, radios, and television sets are the only appliances found in more than half the electrified homes.[58] (See Table 6.)

The impact of electricity varies in different villages, depending on the society and the way in which power is introduced. However, studies

"In most villages people believe that electricity improves their standard of living more than any other change they have experienced."

Table 6: Use in Homes in Rural Costa Rica and Colombia[1]

| | Share of Homes | |
Appliance	Costa Rica	Colombia
	(percent)	
Lights	100	95
Iron	57	73
Television	54	39
Radio	52	79
Refrigerator	42	19
Blender	39	32
Washing machine	31	—
Electric stove	21	91
Sewing machine	3	26
Fan	—	7

[1]Based on household surveys.

Sources: Randy Girer, "Rural Electrification in Costa Rica"; Eduardo Velez, "Rural Electrification in Colombia."

show that in most villages people believe that electricity improves their standard of living more than any other change they have experienced. Women appear to appreciate the benefits of electricity more than men, since they generally spend more time around the home and electricity can help in household chores, while fans and radios make leisure time more pleasant. Many women report that they have more free time after getting electricity. Frequently, electric pumps are used to provide a reliable, clean supply of water from a village well for the first time, which makes life easier and improves health.[59]

But even at their best, Third World rural electric systems rarely match the reliability taken for granted in industrial countries. Wide voltage fluctuations make it impossible to use some appliances, and blackouts are common. Often, people must go for days without power, waiting for a spare transformer to be carried in over rutted roads or for repairs to be made to storm-damaged lines. Since line crews are often inexperienced and working on live wires can be dangerous, managers frequently shut down an entire system to make minor repairs.

Planners aim to achieve many additional benefits through rural electrification. Several studies have found that electricity promotes literacy by making it possible for children to study in the evening. In rural India, television-viewing has become widespread, thanks both to electricity and a new satellite that transmits national programming into remote areas. The government broadcasts educational programs, including lessons in preventive health care and farming techniques.[60]

Electricity has also been proposed as a way to reduce urban migration by making village life more attractive, but studies show that it may have the opposite effect by exposing people to the "charms" of urban life through television. Proponents also suggest that rural electrification can help reduce birth rates by providing alternative evening activities, but there is no persuasive evidence that this is the case. However, sometimes electricity provides unexpected benefits. In a remote village in China's Fujian province in which young men have traditionally had a hard time finding wives, the arrival of electricity has attracted more brides.[61]

Rural electrification can have the unfortunate effect of perpetuating inequality, particularly in poor villages where only wealthy households can afford power. In the Philippines for example, newly electrified villages generally have a hook-up charge of $30 to $40, which is beyond the means of many poor families. A survey in India found that poor families took an average of ten years after initial electrification before they could afford to connect to the system. In Costa Rica, preferential hook-up rates and lifeline electricity prices for the poor have resulted in most of the homes in electrified areas receiving power, but even there, poor households use a small fraction of the electricity that wealthy households do. Special efforts must be made continually if the benefits of electricity are to be widely shared.[62]

In most villages, even the best electrification program can only address part of the energy problem. The main use of energy in rural areas is for cooking, a task that often is done with wood, crop residues, or dung. These biomass sources provide an estimated 48 percent of all the energy used in the Third World, and in many rural households the figure is over 90 percent. Because of the lower efficiency of these biomass energy systems, cooking requires approximately five times as much energy as it does in cities.[63]

A survey of two electrified Indian villages shows that even in the wealthier of the two, fuelwood provides 68 percent of the energy and electricity just 3 percent. (See Table 7.) There is a steadily worsening shortage of cooking fuels that is forcing villagers in many areas to walk long distances to obtain firewood. In many circumstances, rural electrification has done little to ease wood shortages. A project in Nepal has been criticized because the electric light it provides encourages villagers to stay up late at night and burn more of the area's dwindling firewood.[64]

Electricity supplies just 4 percent of the Third World's energy, and very little electricity is used for cooking. Even in electrified homes, most village families cook with biomass. The reason is that heating food requires more electricity than most people can afford. (A simple hot plate can use over 500 watts of power, or 33 times as much as the 15-watt lightbulb that is the main source of light in many village homes.) Only in areas with abundant hydropower (such as Colombia) is electricity used for cooking.[65]

Electricity is simply not the best way of solving the cooking-fuel problem in most villages. A variety of other initiatives must accompany electrification programs in order to effectively address the broad range of rural energy problems. Promising approaches that have been tested successfully in some countries include fuelwood plantations, more efficient cookstoves, biogas digesters (to turn biological wastes into flammable methane gas), and solar ovens. Although these technologies have been extensively tested and widely disseminated in

Table 7: Energy Use in Two Indian Villages

40

Activities	Energy Consumption	
	Village A	Village B
	(percent)	
Cooking	64	91
Agriculture	22	3
Industry	7	4
Lighting	4	2
Transportation	3	—

Sources

	(percent)	
Fuelwood	68	89
Human/Animal Power	24	89
Kerosene	5	2
Electricity	3	1

Sources: Roger Revelle, "Energy Use in Rural India," *Science*, Vol. 192, June 1976; N. H. Ravindranath et al., "The Design of a Rural Energy Center for Pura Village," *ASTRA*, Indian Institute of Science, 1979.

some countries, they have received far less funding than rural electrification has.

Other important rural energy uses include water pumping, grain grinding, and many other mechanical processes. These can be performed with electrically driven motors, but they can also be powered by biomass or human or animal power, particularly if technologies to improve their effectiveness are developed. However, some applications of electricity have no good substitute. These include high-quality lighting, refrigeration, television, computers, and telecommunications. The question is whether by themselves these uses are great enough to justify traditional rural electrification by extending the central grid. In many cases the answer may be no. If

"The real potential of electricity lies not in
providing social amenities but in
stimulating long-term economic
development."

consumers can only afford to use electricity for lighting and television, then rechargeable batteries or small diesel generators might be
sufficient and far less costly. There is a need for comprehensive rural
energy planning and the introduction of a broad range of appropriate
technologies. While electricity can be beneficial, it should not get
exclusive priority.

The real potential of electricity lies not in providing social amenities
but in stimulating long-term economic development. One way this
can happen is through the creation of rural industries. So far the
record is mixed. Where economic expansion has already begun and
markets are developing for various products, the development of
rural industries has followed electrification. But in poorer, more remote areas, factories have not flourished after the arrival of electricity.

In Indonesia, electricity has encouraged household-based cottage industries, owing in part to the availability of lifeline prices, but there
has been little development of larger rural factories that must pay
higher prices. Costa Rica has seen some use of electricity in sawmills,
cement factories, and tourist hotels, but electrification has not been a
major boost to rural industry. Electricity in Colombia has encouraged
the development of small businesses, particularly roadside stores.
But in most countries, development planners find electricity is not
enough. Rural industries require a variety of other infrastructure
investments in roads, training, and financial credit.[66]

Electricity is also intended to raise agricultural productivity, but so far
farm use of power is mainly limited to large plantations owned by the
wealthy. For typical small farms, there are just not many useful,
affordable ways to employ electricity, though limited on-farm food
processing is practiced in some areas. In Costa Rica, electricity is
rarely used in the production of coffee, sugar, or vegetables, but it has
helped raise the output of poultry and dairy farms. In Bolivia, planners had hoped that electricity would encourage widespread irrigation, but the harsh climate, salty soils, and lack of financial support
cut these plans short.[67]

Rural electrification in India and Pakistan is largely directed to developing irrigation in semi-arid parts of the country. This has been quite successful though at substantial cost. One study found that Indian farmers were only paying between one-fifteenth and one-tenth of the actual cost of providing them with electricity. Irrigation has grown rapidly in newly electrified areas, and this, together with the introduction of improved plant varieties and fertilizer, has raised agricultural yields. It is now possible to pump groundwater to grow wheat, sorghum, and other grains in areas with extended dry seasons, benefiting some of the poorer areas of India. However, in Pakistan where nearly one quarter of the country's power is used in irrigation, waterlogging and salinity have undermined the productivity of large areas of irrigated land.[68]

Given the many needs of rural people, investment priorities, including electrification, must be balanced. Electricity can help to meet basic needs, but it may not be the fastest or most cost-effective way to do so in sparsely populated areas or regions with mountainous terrain. The Papua New Guinea "Energy White Paper," published in 1979, said, "It is clear that investment in rural electrification is investment foregone in such areas as improved roads, water supply, schooling, and health services. When given a choice of these alternatives, in the face of the real costs and benefits of each, it is debatable whether rural electrification would rank high on a list of village priorities."[69]

Rural electrification is not a blanket cure for the ills of village life. Rather, it is one tool that is appropriate in some cases. Since village conditions vary greatly, careful studies should be done before choosing a package of development tools. Health care and simple farming technologies may deserve priority. Electrification should be linked to other appropriate energy and development programs such as the introduction of efficient cookstoves.

Electric cooperatives offer an approach to rural electrification that has worked well in some countries. Electric co-ops have their roots in the U.S. rural electrification program of the thirties. An electric cooperative is a local agency, owned by its members and managed by a local staff, but often supported financially by a national rural electrification administration. (It is usually expected that the co-op will repay the initial loan for rural electrification over a period of 20 to 30 years.) Many rural electric co-ops have successfully overcome the indifference of central utilities to the needs of rural areas, but in practice they usually do not attain full independence from government planners. Most co-ops do not generate their own power; rather they purchase it at a wholesale or subsidized price from the national power authority.[70]

Several developing countries have established electric cooperatives since the sixties, but overall they are responsible for a small share of rural electricity systems worldwide. Some of the earliest co-ops are in Costa Rica, where three were set up with the help of the U.S. Agency for International Development in the early sixties. Subsequent evaluations show that these co-ops have been quite successful. They now provide electricity to about half of the families in their service areas, and they have largely repaid their initial debts. The areas where the co-ops are located have experienced substantial economic growth.[71]

Importantly, electric cooperatives can play a crucial role in encouraging productive uses of electricity. In the many areas where electrification has not led to improvements in agricultural productivity or the development of rural industries, one of the main reasons is a lack of technical skills or financial resources. Co-ops can be organized to provide low-interest loans and technical extension services for everything from small lumber mills to irrigation systems. Electricity for its own sake brings limited benefits. Locally based, community-owned organizations are more likely to encourage and support income-generating industries than are national power authorities.

Electric co-ops are often initially vibrant, democratic organizations in which member participation is high. In some villages, electric cooperatives have provided a unique opportunity for citizen involvement. Studies show that when local people are involved, there is less stealing of electricity, equipment is better maintained, and productive uses of electricity are likely to develop more rapidly. Over the years, interest and participation in co-op programs usually wane, and management is left to the executives and board members. Nonetheless, electric co-ops can and do serve as catalysts for locally based development initiatives.[72]

The importance of local involvement is seen in the Philippines, where a cooperative-based rural electrification program has provided power to more than 18,000 villages in just 12 years. Annual meetings of the electric co-ops have large turnouts and boisterous festivities. Many electric co-ops in the Philippines provide much more than electricity. They have become involved in charcoal production, school lighting, school waterworks, and vegetable gardening.[73]

In the Philippines and elsewhere, the effectiveness of electric cooperatives varies widely depending on the quality of local leadership. If the cooperative becomes a pawn of politicians, it is often ineffective, but if it can be insulated from political pressures, the chances of its success are much higher.

Among the countries now looking seriously at the potential to expand cooperative-based electrification are Bangladesh, India, and Indonesia. All three have experience with electrification programs that rely largely on the central utilities, and they are considering a change in direction. Many of the programs have required large subsidies and yielded small benefits. Given their current fiscal crises, these countries are hoping to find more effective and less expensive approaches to electrification. In India, a few small electric cooperatives organized with U.S. assistance in the sixties are now serving as models. And in Indonesia, electric co-ops set up largely at local initiative during the seventies are being evaluated.[74]

The most important role that electric cooperatives can play is in decentralizing the decision-making process and bringing local people into comprehensive energy and development planning. Indeed, rural electrification outside the context of an overall development strategy is bound to fail. What is really needed are broader energy and development cooperatives that can also address the cooking fuel issue—the central energy problem in most rural areas. Co-ops can effectively promote fuelwood plantations, more efficient cookstoves, and a range of other development programs. Electricity may have a greater impact if development agencies consider it as part of a larger package of initiatives for rural areas, rather than as an end in itself.

Decentralizing Generators and Institutions

Utility planners in industrial and developing countries often share a bias: They assume the ideal generator size to be extra large, and the most efficient utility to be a government- or privately owned monopoly. The history of power systems is littered with small industrial power generators that have been abandoned and tiny municipal and private utilties that have been swallowed up by national power authorities. No doubt this trend has often led to less expensive and more reliable electric service, but at times it has not. Recent technological developments should encourage utility planners to look more closely at small decentralized power generators—particularly for rural electrification.

The relatively limited rural electrification that has occurred so far has concentrated in areas where grid extension is relatively easy and economical—heavily populated agricultural plains and prosperous areas surrounding major urban and industrial centers. Left behind are remote, forested and mountainous regions where populations are sparse and cash incomes are limited or nonexistent. Examples include much of the Andes region of South America, most of Central America, large sections of India, Indonesia, and Pakistan, and virtually all of Afghanistan, Nepal, Bhutan, Burma, and Laos.

In addition, many island communities will never be reached by grid electricity. The Philippines, for example, consists of over 7,000 individual islands, and even when the national grid is fully completed, many villages will not be connected. Full rural electrification, if that is the goal, can never be completed by central grids alone. Other areas with extensive islands include Indonesia, large stretches of the South Pacific, and the Caribbean.

The next frontier in rural electrification and in rural energy development generally is the use of decentralized power technologies relying on renewable resources. Although these are particularly important in more remote areas, they are also likely to prove useful for supplying small towns, industries, and even the central power grids of many developing countries. Whereas a 100-megawatt power plant is considered small in most industrial countries today—one-tenth the size of a standard nuclear unit—in many developing countries it is considered large. Many Third World power systems rely primarily on small diesel generators and oil-fired steam plants rather than the large coal and nuclear units that dominate the power systems of industrial countries.

One of the chief advantages of small power projects is that they can be built quickly. At a time when demand forecasts and future economic conditions are uncertain, small projects can be planned to match actual needs rather than dubious forecasts. Financing charges typically account for a large share of the expense of power projects, which can be greatly reduced if a project is completed quickly. The real cost of borrowing money today is substantially higher in most Third World countries than in the industrial world, which makes small, short lead-time projects even more attractive. In addition, small power projects often have lower foreign exchange requirements than larger ones since more of the parts can be built locally.

Diesel generators have been used for small-scale power generation in remote areas for decades, and provide important experience in the operation of decentralized systems. An estimated 580 megawatts of stand-alone diesel sytems are now in use in the Third World. Diesel generators of 10 kilowatts or less are used to power individual homes and communications systems, while larger generators of several hundred kilowatts or more can power a whole village or military encampment.[75]

"The next frontier in rural electrification
and in rural energy development generally
is the use of decentralized power
technologies relying on renewable
resources."

Diesel generators are relatively inexpensive to install—$400 to $1,000 per kilowatt, or less than $10,000 for a 10-kilowatt system—but expensive to operate. Diesel generators are relatively inefficient and even with lower oil prices the cost of fueling them is substantial. Total diesel generation costs typically range from 15¢ to 50¢ per kilowatt-hour, compared to an average cost of grid electricity in the Third World of about 7¢ per kilowatt-hour. These high operating costs mean that most villages that rely on diesel generators can operate the system for only a few hours a day, usually in the late afternoon and early evening when food is prepared and lighting is needed.[76]

Other disadvantages of diesel generators include a need for frequent maintenance; neglect often leads to breakdowns and time-consuming repairs. Often a damaged generator must be fixed in a distant town. A village desiring reliable power service has to have a spare generator ready if the primary system breaks down, which adds further to costs. In addition, the diesel fuel must often be carried long distances over unpaved roads and is sometimes not available after a heavy storm or if the government should restrict imported fuel because of foreign exchange shortages. For all their disadvantages, diesel generators have served an important role in many situations where no other technology would do. Today, however, there is a much wider array of small power options available, many of which can substitute for or complement diesel systems.

Small hydropower is the most mature small power alternative. Usually classified as less than 15 megawatts in capacity, small hydro's advantage is that the technology is well developed and economical. A study for the U.S. Agency for International Development found that 10,000 megawatts of non-grid-connected small hydro projects had been constructed in the Third World by 1983. The development of small hydro has only begun, however, and many developing countries are now actively exploiting their hydropower potential. One study projects that the capacity of stand-alone small hydro systems will reach 29,000 megawatts in 1991. Total worldwide potential is probably well over 100,000 megawatts. Detailed surveys have been carried out in few countries, but preliminary estimates indicate that

48 small hydropower potential in many countries exceeds total installed power capacity.[77] (See Table 8.)

Despite growing interest in small hydropower, it receives a small share of total power funding. There are several reasons. Few countries have done a thorough inventory of small hydro sites and planners do not know the potential. Many nations lack the technical expertise to design and build small hydro projects. And many Third World utilities have a continuing bias toward large projects managed by multinational firms; small projects designed and built locally are rarely considered.

One of the difficulties in small hydropower development is that each project is unique, dependent on the topography and hydrology of the site. Each requires special design and construction considerations that would be unnecessary for a diesel generator. Skilled engineers and managers must be available to assess the feasibility of a proposed project and then carry it through. Often, engineers trained in the West overdesign small hydro projects, using the specifications employed in industrial countries. Their designs lead to the need for extensive foreign consultation, the import of many devices and materials, and high costs. Many of the smaller hydro projects are simply infeasible if this approach is employed.[78]

Since 1980, substantial efforts have been made to develop indigenous, appropriate approaches to small hydropower development. Some have been started by government and multinational agencies, but others are the result of village-level experimentation. Many micro-hydro projects under 100 kilowatts in capacity are "run-of-the-river" systems that require less extensive civil works and employ locally manufactured equipment. Experience with small mills and irrigation systems often provides a pool of local talent to carry out such projects.[79]

As small hydro projects multiply, entrepreneurs in India, Nepal, and Pakistan have started manufacturing pipes, turbines, generators, and load controllers locally. Similar enterprises have begun to appear in

Table 8: Decentralized Hydropower Potential in Selected Countries

Country	Small Hydro Potential	1982 Installed Power Capacity[1]	Small Hydro Potential as Share 1982 Capacity
	(megawatts)	(megawatts)	(percent)
Peru	12,000	3,300	360
India	10,000	35,400	28
Philippines	4,000	4,800	83
Costa Rica	2,700	650	415
Thailand	1,100	4,630	22
Indonesia	1,000	5,170	19
Guatemala	1,000	530	190
Nepal	800	138	580
Guinea	560	102	550
Bolivia	500	500	100
Pakistan	300	4,100	7
Madagascar	270	190	142
Sri Lanka	200	500	40
Liberia	150	360	42
Jamaica	100	680	15

[1]Figure reflects country's total generating capacity, not just hydropower.

Source: Hagler, Bailly & Company, "Decentralized Hydropower in AID's Development Assistance Program."

parts of Africa and Latin America where small hydro development is active. Such innovations have reduced the cost of some small hydro projects from the $3,000 to $6,000 per kilowatt typical of many heavily engineered systems to as low as $1,000 to $2,000 per kilowatt.[80]

Nepal, located on the southern edge of the Himalayas, is one of the most difficult countries to supply with a central power grid and one of the best endowed with small hydropower potential. Both the Ne-

50 palese government and international development agencies have undertaken small hydro development in Nepal, but the most successful efforts are low-budget projects constructed by private nonprofit organizations. The United Mission to Nepal, an indigenous, church-supported organization, has helped establish a private small hydro industry that has installed 65 water-powered mills since the mid-seventies, most of them for grinding grain in areas that are more than a day's walk from the nearest road. Most of the components are manufactured in Nepal, and many innovations have been made as development has proceeded. Although most of these projects are mechanical mills used to grind grain or saw logs, many of the lessons can be applied to water-powered electricity generators. Similar efforts have succeeded on a smaller scale in India, Pakistan, Zaire and elsewhere.[81]

Isolated small hydro projects, many of them supported by aid agencies, have appeared throughout the world. A 12-kilowatt plant in Khun Khong, Thailand, generates electricity for a village and adjacent forestry station. A larger 3,000-kilowatt plant provides power for a nearby town. On the island of Bali in Indonesia, a generator has been added to an irrigation project, providing 90 kilowatts of power to a nearby town. In Chongos Alto, Peru, villagers did the excavation for a project that will provide them with 425 kilowatts of electricity.[82]

Local participation and investment is the key to the world's most extensive small hydropower program in China. China now has 76,000 hydro plants that supply 9,500 megawatts of power. Most of these plants have been built since 1970 and are the foundation of what the World Bank terms, "the most massive rural electrification effort ever attempted in the developing world." About 40 percent of China's rural townships and one third of its 2,200 counties get most of their power from small hydrodams.[83]

Although overall plans are developed by China's leadership, "local units" ranging from counties to towns, brigades, and communes are in charge of much of the planning, execution, and financing of the country's small hydro projects. The national government has focused most of its effort on providing technical support. Large subsidies were once available from the national government but most were phased out in 1985. Many small hydro projects started as autonomous systems, but were later linked together in local grids or connected to larger central grids. China has effectively shown how to build power systems from the bottom up.[84]

China has also demonstrated that given the necessary incentive, spirit of cooperation, and local leadership, power projects can be carried out successfully at the local level. County officials are instructed to rely on the "three selfs"—self-construction, self-management, and self-consumption. The small hydropower program appears to be thriving even in the midst of the country's new economic policies which rely much more heavily on free markets and private initiative. For more than 100 Chinese hydro equipment manufacturers, business is booming. China recently targeted 100 additional counties for electrification using small hydropower by 1990 and the government plans to more than double total small hydro generation by the year 2000.[85]

Although China's approach is not applicable in the last detail to other countries, the general model is. The blend of local and national involvement, and private initiative and public cooperation has proved effective. It is now apparent that neither a completely government-dominated power system nor an entirely deregulated private approach can capture the ideal mix of economic efficiency and reliable service.

Efforts are also under way to develop small-scale power systems using biomass energy sources such as wood and agricultural wastes. The advantage of biomass is that the fuels can be stored for use when needed. Despite the fuelwood shortage in some regions, there are many rural areas that have ample supplies of under-utilized biomass, such as rice hulls or coconut shells. In addition, special crop or forest plantations can be established to grow biomass specifically for use in power generation.

Biomass-conversion technologies are more complex than hydro generators. Among the technologies used, the simplest is direct combustion of biomass, using the heat to run a reciprocating steam engine or a steam turbine, either of which can be connected to a generator. Another approach is to gasify the biomass, using the gas to run a diesel or internal combustion engine. The village of Picon in Indonesia has installed two 40-kilowatt wood gasifiers, using the electricity for pumped irrigation, food processing, and woodworking. Finally, some biomass sources can be converted to liquid fuels such as ethanol or coconut oil. In the Philippines, some small generators are run on coco-diesel, a mixture of diesel fuel and coconut oil.[86]

The Philippines is also the site of the only large-scale effort to supply a central power grid with electricity from wood-fired plants. The National Electrification Administration started this "dendrothermal" program in the late seventies to help reduce the country's costly dependence on imported oil. Local labor and management developed plantations of fast-growing leucaena trees for fuelwood. Equipment for 17 power plants was imported from Britain and France.

Like many crash projects, the dendrothermal program ran into trouble. Some of the plantations were started on extremely poor soil and failed; much of the generating equipment required extensive repairs before it would work properly; and the country's worsening economy forced a cutoff of funds for rehabilitating the program. Only three of the dendrothermal plants were working in 1985. This is unfortunate because most of the problems appear to be resolvable, and economic projections show that the dendrothermal plants can produce power for less than the average cost of generation in the Philippines. Too often, planners undertake innovative projects like this one but lack the fortitude to work through the start-up problems that are inevitable with first-time projects.[87]

Wind power also has considerable potential in some developing countries. Employed widely by American midwestern farmers in the twenties before rural electrification, wind generators are proving effective in similar settings in the Third World. The most extensive effort so far is on the remote windswept plains of Inner Mongolia, where Nomadic herdsmen are using 2,000 small wind turbines for lighting, television, electrifying corral fences, and projecting movies. A portable wind turbine has been designed so that the nomads can carry their power source with them. Three Chinese factories are now producing several thousand wind generators each year for use in Tibet, Xinjiang, and other remote areas.[88]

Grid-connected collections of wind generators called "windfarms" also show promise. Windfarms were first built in California in the early eighties and now supply the state with about 1,100 megawatts of power. Utility-sponsored studies show that the better windfarms can produce power at a cost of about 7¢ per kilowatt-hour, which is competitive with conventional power sources in the United States. Several developing countries are now studying the potential of wind-farms, and small experimental projects are being tested in China, India, and Pakistan. European and American firms have signed agreements to establish joint wind turbine manufacturing facilities in each of these countries. Third World wind-power development is still at an early stage, and there is a need for detailed wind assessments and feasibility studies. Early evidence indicates that wind power will soon take its place as a decentralized power source that is economical in many areas.[89]

Solar photovoltaic cells are in a sense the ultimate decentralized power source since they rely on sunlight, a more widely available resource than wind, biomass, or falling water. Moreover, solar cells directly produce electricity, requiring no generator or extensive civil works. However, if electricity is needed after dark or in cloudy weather, storage batteries or a backup generator must be added to a photovoltaic system which can double the cost. Even without storage, photovoltaics can provide important amenities to villages that cannot afford other alternatives. The main barrier to photovoltaics so far is the relatively high cost—about $10,000 per kilowatt or a generating cost of over 50¢ per kilowatt-hour. Development of solar cells is progressing rapidly however, and many experts expect their cost to be cut at least 50 percent during the next decade.[90]

Even today, for some very small uses of electricity in remote areas (less than one kilowatt) photovoltaics are less expensive than a diesel generator. Remote communications systems and weather stations now commonly use photovoltaics, and there are several thousand remote homes in the United States that have solar power systems. Foreign aid programs have brought solar power systems to a number of Third World villages. In 1978, the U.S. Agency for International Development installed a 1.8-kilowatt photovoltaic system at the Tangaye village in Burkina Faso that is used mainly for lighting, water pumping, and grain grinding. A much larger 25-kilowatt system was installed in a Tunisian village in 1982, also for water pumping and household tasks. A West German team installed another 25-kilowatt system in an Indonesian fishing village in 1984 to produce ice for storing the fish catch.[91]

The largest village photovoltaic project to date is on the island of Faaite, Tahiti, where over 550 homes were electrified in 1982 and 1983, using solar cells manufactured in France and systems designed and built in Tahiti. With the help of government subsidies, the systems were quickly installed on rooftops at a cost of $1,500 per house. This project was carried out by a private French Polynesian firm that hopes to install similar systems in other parts of the South Pacific.[92]

Third World interest in photovoltaics is rising quickly. Several countries, including China and India, have recently signed agreements with foreign companies to begin the domestic manufacture of photovoltaic systems. "Turnkey" photovoltaic plants can now be established almost anywhere in the world, and developing countries can make a gradual transition from simply assembling systems locally to developing fully integrated domestic manufacturing processes. Studies in India show that photovoltaics are now economical for many small applications, and Indian companies are installing several hundred systems each year. If costs continue to fall as projected, there seems little doubt that photovoltaics will eventually become the most important power source in many of the world's villages.[93]

"If costs continue to fall as projected, there seems little doubt that photovoltaics will eventually become the most important power source in many of the world's villages."

Companies in the United States are now building and marketing small hybrid power systems that incorporate wind or solar generation with a diesel generator for backup power. The North Wind Company in Vermont has installed 125 hybrid systems in 17 countries, most of them incorporating wind power or photovoltaics. Another company, Earth Energy Systems of Minnesota, sells a 10-kilowatt system that includes photovoltaics, a wind turbine, diesel generator, and batteries. Although it costs at least $7,000 per kilowatt of total capacity—far more than most central systems—in many remote areas it is cheaper than the alternatives. The key to the economics of such systems is that their size be carefully matched to power needs and that the power not be used wastefully.[94]

It is notable that little of the development of decentralized power systems has been the work of national power authorities. They are simply not accustomed to executing small projects at the local level. This opens a wider issue: Should the private sector be brought into the power generation business? This is a promising concept, but runs against the grain of power-planning over the past two decades. In Latin America in particular, long battles were waged to abolish private utilities, and factories that cogenerate power were forced to shut down. But current proposals to introduce competition into the power industry are not aimed simply at bringing back the private utility monopolies of the fifties. Rather, the idea is to create hybrid competitive power systems in which there is some central planning and control, but in which private companies are encouraged to bring innovation and efficiency to the system.

Recent initiatives in opening up the electricity system to competition have come in countries where power expansion programs have run into significant problems. These include Colombia, Turkey, and Pakistan. The utilities in these countries have reached the limit of their capacity to finance and manage additional projects, and will experience power shortages if construction budgets continue to be cut. In Pakistan, the Water and Power Development Authority is already short of capacity by 1,200 megawatts, and rolling blackouts are common.[95]

The World Bank and the U.S. Agency for International Development are actively studying the feasibility of private power projects in several countries. Private companies, probably in partnership with foreign corporations, would be offered contracts to build and operate power plants and sell the power to the national utility at a fixed price. The foreign partners would help raise the foreign exchange needed and could own equity in the project. Both large central power projects and small decentralized facilities could be built independently of national power authorities.

Private power projects have flourished in the United States in the early eighties largely as a result of the Public Utility Regulatory Policies Act (PURPA) passed in 1978. Under the act, U.S. utilities are required to purchase power from independent producers at the long-run marginal cost of generation. This, together with the availability of tax credits, has resulted in over 25,000 megawatts of proposed small power projects, half of which is fossil-fuel-fired cogeneration (combined production of heat and power) and the rest a diverse mix of wind power, hydropower, wood-fired plants, waste-fired plants, and solar power projects. In California, enough power contracts have been signed to provide more than one third of the state's electricity.[96]

Several developing countries are now studying the U.S. model and privately developed generating plants are beginning to appear in some countries. In the state of Gujarat, India, the State Electricity Board offered in 1985 to pay the equivalent of 10¢ per kilowatt-hour for privately generated wind power. It is too early to know the outcome of this offer or whether it will be extended to other energy sources, but Gujarat already has the Third World's first operating windfarm, and its state government has successfully supported the development of other renewable energy sources. The Chinese government is encouraging local governments and industrial enterprises to pool their resources to build power plants independently of the national Ministry of Water Resources and Electric Power.[97]

Opening the power system to private competition is both a way to bring financial resources to electricity projects and a way to spur the development of innovative technologies. Countries with different political systems and power industries seem to be moving in similar directions, which makes the new developments particularly noteworthy. The end of the era of exclusive government power monopolies will almost certainly open opportunities to improve the reliability and cost-effectiveness of electricity systems.

There are no simple prescriptions to solve the Third World's power problems. The needs are vast and the available resources small. New initiatives must be carefully designed if they are not to create as many problems as they solve. However, some risks must be taken because existing institutions are foundering beneath the weight of the problems that confront them.

Increased efficiency, more attention to the special energy needs of villagers, and the development of decentralized power technologies can together contribute significantly to the viability and effectiveness of Third World power systems. However, the most fundamental change needed is a philosophical one. Electricity should not be considered an end in itself but rather a means to reaching broader development goals. Less effort should be placed on accelerating the pace of expansion of power systems and more on ensuring that they effectively promote development.

Notes

1. A. Heron, "Financing Electric Power in Developing Countries," *IAEA Bulletin* , Winter 1985; Edward S. Cassedy and Peter M. Meier, "Planning for Electric Power in Developing Countries in the Face of Change," draft chapter for *Planning For Changing Energy Conditions* (New Brunswick, N.J.: Transaction Inc., forthcoming).

2. 1985 estimates based on figures in World Bank, "1982 Power/Energy Data Sheets for 104 Developing Countries," Washington, D.C., March 1986; World Bank, *China: The Energy Sector* (Washington, D.C.: 1985); and Edison Electric Institute, "Electric Output," Washington, D.C., April 30, 1986.

3. Mohan Munasinghe, World Bank, private communication, May 19, 1986.

4. Worldwatch Institute estimate based on figures in several regional and country studies. In China, half of the rural households, representing 400 million people, do not have electricity. In other developing countries an average of 75 percent of the 1.7 billion rural people do not have power. Even in "electrified" villages, many homes lack power.

5. Historical material from Hugh Collier, *Developing Electric Power: Thirty Years of World Bank Experience* (Baltimore, Md.: Johns Hopkins University Press, 1984).

6. Ibid.

7. Ibid.; The World Bank, "Summary of FY80-85 Bank Power Lending," unpublished memorandum, Washington, D.C., August 1, 1985.

8. Inter-American Development Bank, *Inter-American Development Bank Annual Report 1985* (Washington, D.C.: 1985); Asian Development Bank, *Annual Report 1985* (Manila, Philippines: 1985); African Development Bank, *The African Development Bank 1964-1984* (Abidjan, Ivory Coast: 1984).

9. Worldwatch Institute estimate based on various sources. Installed capacity of 450,000 megawatts at a cost of $1.1 million per megawatt would require an investment of $500 billion.

10. Collier, *Developing Electric Power*; Renewable Energy Institute, "India: An Overview of the Electric Power Sector and its Socio-Economic Environment," unpublished, Washington, D.C., 1985.

11. Collier, *Developing Electric Power*.

12. Cassedy and Meier, "Planning for Electric Power in Developing Countries in the Face of Change."

13. World Bank, "1982 Power/Energy Data Sheets for 104 Developing Countries."

14. Ibid.

15. Heron, "Financing Electric Power in Developing Countries."

16. World Energy Conference, *Survey of Energy Resources, 1980* (Munich: 1980).

17. Figures compiled by Worldwatch Institute from various sources. See Lester R. Brown et al., *State of the World 1985* (New York: W.W. Norton, 1985).

18. World Bank, *China: The Energy Sector.*

19. Patricia Adams and Lawrence Solomon, *In the Name of Progress: The Underside of Foreign Aid* (Toronto: Energy Probe Research Foundation, 1985).

20. Ken Lieberthal, "Energy Decision-Making in China," lecture at the Johns Hopkins School of Advanced International Studies, Washington, D.C., February 5, 1986.

21. World Bank, *China: The Energy Sector.*

22. Ibid.; Massachusetts Division of Air Quality Control, "Acid Rain and Related Air Pollution Damage: A National and International Call for Action," unpublished, Boston, Mass., August 1984.

23. World Bank, *The Energy Transition in Developing Countries* (Washington, D.C.: 1983).

24. James Everett Katz, *Nuclear Power in Developing Countries* (Lexington, Mass.: Lexington Books, 1982).

25. *Nucleonics Week*, January 30, 1986.

26. Jasper Becker, "China Switches from Nuclear Power to Hydroelectricity," *New Scientist*, April 3, 1986.

27. Philippines Ministry of Energy, *1984 Annual Report* (Manila: 1984).

28. Collier, *Developing Electric Power*.

29. Heron, "Financing Electric Power in Developing Countries."

30. Worldwatch Institute estimate, based on figures available for selected countries; National Rural Electric Cooperative Association (NRECA), "Central America Rural Electrification Study," unpublished, Washington, D.C., 1985.

31. Heron, "Financing Electric Power in Developing Countries"; World Bank, *The Energy Transition in Developing Countries*.

32. Robert Ichord, U.S. Agency for International Development (AID), talk at the Society for International Development, September 12, 1985; "Power Shortage a Priority," *China Daily*, April 29, 1986.

33. World Bank, "Latin America and Caribbean Region Power Sector Finances," unpublished, Washington, D.C., April 22, 1985.

34. NRECA, "Central America Rural Electrification Study"; World Bank, *China: The Energy Sector*.

35. Robert J. Saunders and Karl Jehoutek, "The Electric Power Sector in Developing Countries," *Energy Policy*, August 1985.

36. Fox Butterfield, "Filipinos Say Marcos Was Given Millions for '76 Nuclear Contract," *New York Times*, March 7, 1986,

37. Cassedy and Meier, "Planning for Electric Power in Developing Countries in the Face of Change."

38. U.S. Department of Energy, *Monthly Energy Review*, April 1986; International Energy Agency, *Electricity in IEA Countries* (Paris: 1985).

39. Howard S. Geller, "Progress in the Energy Efficiency of Residential Appliances and Space Conditioning Equipment," in *Energy Sources: Conservation and Renewables* (New York: American Institute of Physics, 1985); Amory Lovins, "Saving Gigabucks with Negawatts," *Public Utilities Fortnightly*, March 21, 1985.

40. World Bank, *Energy in the Developing Countries* (Washington, D.C.: 1980).

41. Jose Goldemberg and Robert H. Williams, "The Economics of Energy Conservation in Developing Countries: A Case Study for the Electrical Sector in Brazil," Princeton University Center for Energy and Environmental Studies, Princeton, N.J., August 1985.

42. Renewable Energy Institute, "India: An Overview of the Electric Power Sector and its Socio-Economic Environment."

43. Howard S. Geller, "End-Use Electricity Conservation: Options for Developing Countries," draft paper for the American Council for an Energy-Efficient Economy, Washington, D.C., March 1986.

44. Howard S. Geller et al., "Electricity Conservation Potential in Brazil," draft paper for the American Council for an Energy-Efficient Economy, Washington, D.C., March 1986. Additional electricity supplies in developing countries cost an average of about $2,000 per kilowatt, including about $500 per kilowatt for transmission and distribution. See Heron, "Financing Electric Power in Developing Countries."

45. Mohan Munasinghe and Jeremy J. Warford, *Electricity Pricing: Theory and Case Studies* (Baltimore, Md.: Johns Hopkins University Press, 1982).

46. Ibid.

47. Geller, "End-Use Electricity Conservation. "

48. Ibid.

49. Howard S. Geller, American Council for an Energy-Efficient Economy, private communication, May 2, 1986.

50. Geller, "End-Use Electricity Conservation."

51. Pacific Gas & Electric Company, *1985 Energy Management and Conservation Activities* (San Francisco: 1984); Nuclear Information Resource Service, Washington, D.C., private communication, May 19, 1986.

52. City of Austin Electric Utility Department, *Austin's Conservation Power Plant* (Austin, Tex.: 1984).

53. World Bank response is quoted in AID, *The Product is Progress: Rural Electrification in Costa Rica* (Washington, D.C.: 1981).

54. Douglas V. Smith et al., "Report of the Regional Rural Electrification Survey to the Asian Development Bank," draft study, Manila, October 1983; Mohan Munasinghe, World Bank, private communication, May 19, 1986.

55. World Bank, "1982 Power/Energy Data Sheets for 104 Developing Countries"; Randy Girer, "Rural Electrification in Costa Rica: Membership Participation and Distribution of Benefits," Masters Thesis for the Graduate Program in Energy, Management and Policy, University of Pennsyslvania, 1986.

56. World Bank, "1982 Power/Energy Data Sheets for 104 Developing Countries"; World Bank, "Electricity Use in India: Third World Rural Electrification Project," Staff Appraisal Report, Washington, D.C., May 7, 1982; World Bank, *China: The Energy Sector*; Smith et al., "Report of the Regional Rural Electrification Survey to the Asian Development Bank."

57. AID, *Rural Electrification: Linkages and Justifications* (Washington, D.C.: 1979).

58. Girer, "Rural Electrification in Costa Rica "; Eduardo Velez, "Rural Electrification in Colombia," Resources for the Future, Washington, D.C., March 1983.

59. Douglas F. Barnes, "Electricity's Effect on Rural Life in Developing Nations," paper prepared for the United Nations University and the International Development Research Center, Ottawa, September 1984.

60. Ibid.

61. Eugene Chang, "Little Plants Give Lots of Power," *China Daily*, December 14, 1985.

62. Douglas F. Barnes, *Electric Power for Rural Growth: How Electricity Affects Rural Life in Developing Countries* (Boulder, Colo.: Westview Press, 1986); Girer, "Rural Electrification in Costa Rica."

63. Alan S. Miller, Irving M. Mintzer, and Sara H. Hoagland, *Growing Power: Bioenergy for Development and Industry* (Washington, D.C.: World Resources

Institute, 1986); Robert H. Williams, "Potential Roles for Bioenergy in an Energy-Efficient World," *Ambio*, Vol. XIV, No. 4-5, 1985.

64

64. Kunda Dixit, "Neon Conquers Nepali Darkness," *Development Forum*, May 1984.

65. Barnes, "Electricity's Effect on Rural Life in Developing Nations."

66. Barnes, *Electric Power for Rural Growth*.

67. Girer, "Rural Electrification in Costa Rica"; AID, *Bolivia: Rural Electrification* (Washington, D.C.: 1980).

68. Barnes, *Electric Power for Rural Growth*; information on Pakistan from Nigel Green, the World Bank, private communication, May 2, 1986.

69. Smith et al., "Report of the Regional Rural Electrification Survey to the Asian Development Bank. "

70. John H. Magill, "Cooperatives in Development: A Review Based on the Experiences of U.S. Cooperative Development Organizations," report prepared for AID, October 1984.

71. AID, *The Product is Progress*.

72. Girer, "Rural Electrification in Costa Rica"; NRECA, "Report on the Philippine Rural Electrification Impact Survey," Washington, D.C., May 1982.

73. Author's observations based on travel in northern Luzon Province, Philippines, with the National Electrification Administration, November 1985.

74. Samuel Bunker, NRECA, private communication, January 24, 1986.

75. "Remote Power Market is Predicted to Swell," *Renewable Energy News*, July 1985.

76. Smith et al., "Report of the Regional Rural Electrification Survey to the Asian Development Bank. "

77. AID, *Decentralized Hydropower in AID's Development Assistance Program* (Washington, D.C.: 1986).

78. NRECA, "Status, Recommendations and Future Directions for the De-centralized Hydropower Program," Washington, D.C., 1983.

79. Allen R. Inversin, "Pakistan: Villager-Implemented Micro-Hydropower Schemes," NRECA, Washington, D.C., 1983.

80. Smith et al., "Report of the Regional Rural Electrification Survey to the Asian Development Bank"; AID, *Decentralized Hydropower in AID's Development Assistance Program*.

81. NRECA, "Nepal: Private-Sector Approach to Implementing Micro-Hydropower Schemes: A Case Study," Washington, D.C., 1982.

82. NRECA, "Small Decentralized Hydropower Program," Washington, D.C., 1985.

83. "Small Power Units Rise," *China Daily*, April 2, 1986; The World Bank, *China: The Energy Sector*.

84. Robert P. Taylor, *Decentralized Renewable Energy Development in China* (Washington, D.C.: The World Bank, 1982).

85. Chang, "Little Plants Give Lots of Power"; He Quan, "Nation Tags 100 Counties for Mini-Hydro Experiment," *China Daily*, April 16, 1986.

86. Miller, Mintzer, and Hoagland, *Growing Power*; Abubakar Lubis, Witono Budiono, and Cahyun Budiono, "Solar Villages in Indonesia," *SunWorld*, Vol. 9, No. 2, 1985; Philippines Ministry of Energy, *1984 Annual Report*.

87. Frank H. Denton, *Wood for Energy and Rural Development: The Philippines Experience* (Manila: Frank H. Denton, 1983); recent appraisals based on author's discussion with Philippines and AID officials in Manila, November 1985.

88. Xu Yuanchao, "Remote Areas Switch on to Windmills," *China Daily*, February 27, 1986.

89. Robert Lynette, "Wind Turbine Performance: An Industry Overview," *Alternative Sources of Energy*, September/October, 1985; Donald Marier, "Developments in Wind Projects Overseas," *Alternative Sources of Energy*, November/December, 1985.

65

90. Christopher Flavin, *Electricity from Sunlight: The Emergence of Photovoltaics* (Golden, Colo.: Solar Energy Research Institute, 1985).

91. Louis Rosenblum et al., "Photovoltaic Systems for Rural Areas of Developing Countries," National Aeronautics and Space Administration, Washington, D.C., 1979; W.J. Bifano et al., "A Photovoltaic Power System in the Remote African Village of Tangaye, Upper Volta," National Aeronautics and Space Administration, Washington, D.C., 1979; "Ambitious 30KWp PV System to be Completed in Tunisian Village," *Photovoltaic Insider's Report*, September 1982; Lubis, Budiono, and Budiono, "Solar Villages in Indonesia."

92. David Hall, "Sunny Island Power," *Development Forum*, May 1984.

93. R. Venkatesan et al., "Cost-Effectiveness of Decentralized Energy Systems," Resources for the Future, Washington, D.C., July 1983; John A. Ashworth, "Renewable Energy Systems Installed in Asia," AID, Washington, D.C., April 1985.

94. Don Best, "Electrifying the World," *Solar Age*, July 1985.

95. Robert Ichord, AID, talk at the Society for International Development, September 12, 1985.

96. Christopher Flavin, "Reforming the Electric Power Industry," in Lester R. Brown et al., *State of the World 1986* (New York: W.W. Norton, 1986).

97. Michael Farmer, GALT Asiatic Corporation, private communication, January 9, 1986; "China Power Sector Decentralized," *China Daily*, June 6, 1985.

CHRISTOPHER FLAVIN is a Senior Researcher with Worldwatch Institute and coauthor of *Renewable Energy: The Power to Choose* (W. W. Norton, Spring 1983). His research deals with renewable energy technologies and policies. He is a graduate of Williams College, where he studied Economics and Biology and participated in the Environmental Studies Program.

THE WORLDWATCH PAPER SERIES

No. of
Copies

1. **The Other Energy Crisis: Firewood** by Erik Eckholm.
2. **The Politics and Responsibility of the North American Breadbasket** by Lester R. Brown.
3. **Women in Politics: A Global Review** by Kathleen Newland.
4. **Energy: The Case for Conservation** by Denis Hayes.
5. **Twenty-two Dimensions of the Population Problem** by Lester R. Brown, Patricia L. McGrath, and Bruce Stokes.
6. **Nuclear Power: The Fifth Horseman** by Denis Hayes.
7. **The Unfinished Assignment: Equal Education for Women** by Patricia L. McGrath.
8. **World Population Trends: Signs of Hope, Signs of Stress** by Lester R. Brown.
9. **The Two Faces of Malnutrition** by Erik Eckholm and Frank Record.
10. **Health: The Family Planning Factor** by Erik Eckholm and Kathleen Newland.
11. **Energy: The Solar Prospect** by Denis Hayes.
12. **Filling The Family Planning Gap** by Bruce Stokes.
13. **Spreading Deserts—The Hand of Man** by Erik Eckholm and Lester R. Brown.
14. **Redefining National Security** by Lester R. Brown.
15. **Energy for Development: Third World Options** by Denis Hayes.
16. **Women and Population Growth: Choice Beyond Childbearing** by Kathleen Newland.
17. **Local Responses to Global Problems: A Key to Meeting Basic Human Needs** by Bruce Stokes.
18. **Cutting Tobacco's Toll** by Erik Eckholm.
19. **The Solar Energy Timetable** by Denis Hayes.
20. **The Global Economic Prospect: New Sources of Economic Stress** by Lester R. Brown.
21. **Soft Technologies, Hard Choices** by Colin Norman.
22. **Disappearing Species: The Social Challenge** by Erik Eckholm.
23. **Repairs, Reuse, Recycling—First Steps Toward a Sustainable Society** by Denis Hayes.
24. **The Worldwide Loss of Cropland** by Lester R. Brown.
25. **Worker Participation—Productivity and the Quality of Work Life** by Bruce Stokes.
26. **Planting for the Future: Forestry for Human Needs** by Erik Eckholm.
27. **Pollution: The Neglected Dimensions** by Denis Hayes.
28. **Global Employment and Economic Justice: The Policy Challenge** by Kathleen Newland.
29. **Resource Trends and Population Policy: A Time for Reassessment** by Lester R. Brown.
30. **The Dispossessed of the Earth: Land Reform and Sustainable Development** by Erik Eckholm.
31. **Knowledge and Power: The Global Research and Development Budget** by Colin Norman.
32. **The Future of the Automobile in an Oil-Short World** by Lester R. Brown, Christopher Flavin, and Colin Norman.
33. **International Migration: The Search for Work** by Kathleen Newland.
34. **Inflation: The Rising Cost of Living on a Small Planet** by Robert Fuller.
35. **Food or Fuel: New Competition for the World's Cropland** by Lester R. Brown.
36. **The Future of Synthetic Materials: The Petroleum Connection** by Christopher Flavin.
37. **Women, Men, and The Division of Labor** by Kathleen Newland.
38. **City Limits: Emerging Constraints on Urban Growth** by Kathleen Newland.

_____ **Total Copies**

Bulk Copies (any combination of titles)
2-5: $3.00 each   6-20: $2.00 each   21 or more: $1.00 each

Calendar Year Subscription (1986 subscription begins wth Paper 68)
U.S. $25.00 _____

Make check payable to Worldwatch Institute
1776 Massachusetts Avenue NW, Washington, D.C. 20036 USA

Enclosed is my check for U.S. $ _____

name

address

city   · **state**   **zip/country**